Food
50

蛆的口水

Maggot Spit

Gunter Pauli

冈特·鲍利 著

田烁 译

丛书编委会

主　任：贾　峰
副主任：何家振　郑立明
委　员：牛玲娟　李原原　李曙东　吴建民　彭　勇
　　　　冯　缨　靳增江

丛书出版委员会

主　任：段学俭
副主任：匡志强　张　蓉
成　员：叶　刚　李晓梅　魏　来　徐雅清　田振军
　　　　蔡雩奇

特别感谢以下热心人士对译稿润色工作的支持：

姜竹青　韩　笑　杨　爽　周依奇　于　哲　阳平坚
李雪红　汪　楠　单　威　查振旺　李海红　姚爱静
朱　国　彭　江　于洪英　隋淑光　严　岷

目录

蛆的口水	4
你知道吗？	22
想一想	26
自己动手！	27
学科知识	28
情感智慧	29
艺术	29
思维拓展	30
动手能力	30
故事灵感来自	31

Contents

Maggot Spit	4
Did you know?	22
Think about it	26
Do it yourself!	27
Academic Knowledge	28
Emotional Intelligence	29
The Arts	29
Systems: Making the Connections	30
Capacity to Implement	30
This fable is inspired by	31

一群鹌鹑正在四处觅食,那块地方满是苍蝇。"我们正在享用最美味的午餐。"一只鹌鹑咯咯笑着说。

A covey of quails is foraging around in an area infested with flies.

"We are having one of the greatest luncheons ever," giggles one quail.

最美味的午餐

Greatest luncheons ever

苍蝇就像是食物工厂

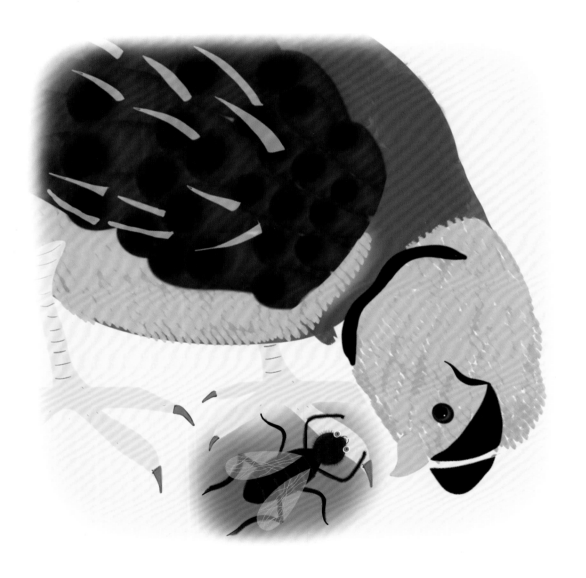

Flies are like food factories

"没错,可是奇怪的是,人类似乎不喜欢苍蝇。"

"是呀,但凡人类自己不懂的,他们就不喜欢。其实,苍蝇就像是食物工厂。"

"同意,对于我们鹌鹑来说的确如此,但人类从来没想过吃蛆吧。"

"嗯,人类很喜欢吃我们小小的、营养丰富的鹌鹑蛋。可当他们看到蛆在垃圾里爬来爬去,就赶紧屏住呼吸,甚至闭上眼睛。"

"Indeed, it is so strange that people simply don't like flies."

"Well, whatever people do not understand, they don't like. Flies are like food factories."

"Yes, for us that is true, but people could never imagine eating maggots."

"Well, they do like our tiny and super nutritious eggs. But when they see maggots crawling in waste, they pull up their noses and even shut their eyes."

"哈，这些胖乎乎的小蛆虫可是我很长时间以来吃过的最美味的东西了。我们还是动作快点吧，要不它们又要变成苍蝇了。说实话，苍蝇可没这么好吃。"

"我同意，但是这些小小的苍蝇卵一旦变成蛆后，它们制造蛋白质的速度可是地球上最快的。你知道吗，只要食物充足，1千克蝇卵只需三天，就可以变成三百多千克富含蛋白质的美味呢！"

"Oh, these chubby maggots are simply the tastiest I have had in a long time. We better hurry or they will turn into flies. And to be honest, flies simply are not as tasty."

"I agree, but there is nothing on earth that makes protein faster than these tiny fly eggs once they turn into maggots. Can you imagine, providing they have enough food, one kilogram of fly eggs turns into more than 300 kilograms of wonderful, protein-rich food within just three days!"

只需三天,富含蛋白质的美味

Protein-rich food within just three days

非常特殊的口水

Very special saliva

"我相信,在这方面没有谁能做得更好,即便是藻类、蘑菇或细菌,都不能和蛆相提并论。你知道蛆有种非常特殊的口水吗?"

"他们的口水能有什么特殊的呢?"

"No one can do better, I believe not even algae, mushrooms or bacteria match this performance! Did you know these maggots have very special saliva?"

"What is so special about their spit?"

"它能帮助治愈伤口。"

"真的吗?怎么做到的?"

"嗯,它不但能清除伤口上死去的组织,还能促进新的细胞生长。"

"It helps heal wounds."

"Really? How does that work?"

"Well, it helps cells to grow while it cleans the wound of dead tissue."

它能帮助治愈伤口

It helps heal wounds

他们为什么不把口水挤出来呢？

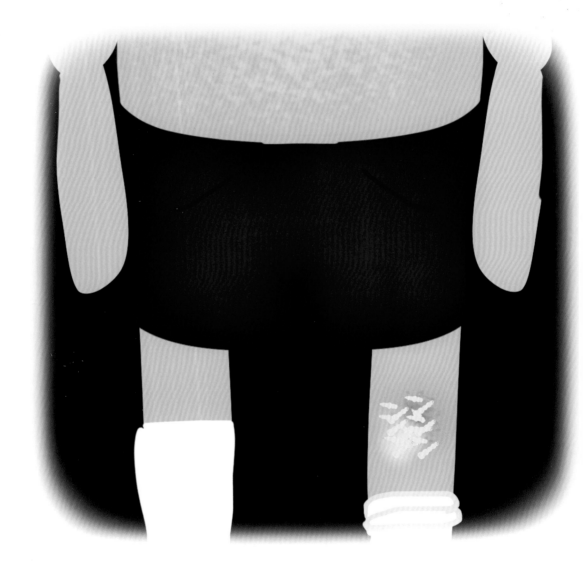

Why don't they milk the spit?

"听起来确实不错,但是你愿意让蛆在你身上爬来爬去吗?"

"如果没有其他选择,我也许会愿意吧。假如伤口不能愈合,那么医生可能不得不给我截肢;这意味着我要失去一只脚或一个翅膀。"

"他们为什么不把口水挤出来呢?"

"That sounds great, but would you like maggots crawling all over your body?"

"Perhaps, if I had no other choice. If a wound does not heal, the doctor may have to amputate; that means losing a foot or a limb."

"Why don't they milk the spit?"

"嗯，提取口水，听起来很有挑战性哦！"

"嘿，如果在海边，你的头扎进水里，想想，会发生什么？"

"Hmmm, extracting spit – now that sounds like a challenge."

"Hey, what happens to you when your head goes under water when you are at the beach?"

听起来很有挑战性哦

Now that sounds like a challenge

啊！你知道我一定会吐出来

Ugh! You know I will throw up

"没什么呀,我会很小心地闭着嘴,屏住呼吸。"

"但是,如果一个巨浪袭来,把你整个儿冲得上上下下,灌进了很多咸咸的海水,这时你会怎么办?"

"啊!你知道我一定会吐出来!"

"Nothing. I take care to close my beak and breathe out through my nose."

"But if a big wave surprises you, turns you upside down and you gulp a lot of salt water, then what?"

"Ugh! You know I will throw up."

"这正是我想的。所以，我们可以把蛆放进盐水中，然后撇去它们吐出的口水。虽然这听起来有些倒胃口，但是的确有助于那些带伤口的人减轻痛苦。"

"你说对了，人类应该意识到，仅靠时间是不能愈合所有伤口的，而蛆虫却能提供一些帮助。"

……这仅仅是开始！……

"That's what I thought. So let us put the maggots in salt water and then skim their spit. It may not sound appetising, but it will certainly provide relief for people with open wounds."

"You are right, people should realise that time alone does not heal all wounds, and that a maggot can certainly offer some help."

... AND IT HAS ONLY JUST BEGUN!...

……这仅仅是开始!……

… AND IT HAS ONLY JUST BEGUN! …

Did You Know?
你知道吗?

Quails have been bred domestically for over 4 000 years and it is thought that the Chinese quail is the ancestor of all breeds of quails. The commercial variety that is most popular originated in Japan.

鹌鹑的人工驯养已经有4000多年历史了,有人认为中国的鹌鹑是所有鹌鹑种类的祖先。那些最受欢迎的商业化品种则起源于日本。

Quails are migratory birds travelling from Africa to Europe and glide using air currents. There are also mountain quails that migrate up and down the mountains on foot.

鹌鹑是一种候鸟,它们在非洲与欧洲之间迁徙,会利用气流来滑翔。还有一类鹌鹑是在山顶与山脚之间迁徙。

When the female has too many eggs and cannot keep them all warm, the male will join her on the nest. The chicks can walk and eat immediately after hatching.

如果雌鸟有很多蛋要孵，以至于不能让所有的蛋都保持适宜的温度，那么雄鸟就会帮着雌鸟在巢中孵卵。雏鸟在孵化后过不多久就会行走和吃东西了。

The Aborigines of Australia, the Mayans in the Andes, and the Italians (during war in the 15th century) used maggots to treat wounds.

澳大利亚土著人、安第斯山的玛雅人以及15世纪经历战乱的意大利人都曾利用蛆来治愈伤口。

Maggots feeding off food scraps and abattoir-waste can produce the same amount of protein in 10 days that a pig generates in six months.

蛆仅靠吃食物残渣在十天中所生产的蛋白质总量，与一头猪半年内的产量是一样的。

The rearing of maggots is highly productive with an annual output of 1.2 tons of black soldier fly larvae per square metre.

养蛆是非常高产的，每平方米年均可生产 1.2 吨黑蝇幼虫。

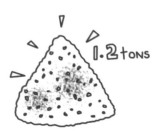

Fly larvae have a good appetite: one kilogram of eggs will convert into more than 300 kilograms of protein in less than a week, provided that there is enough food.

苍蝇幼虫胃口很好，如果食物充足，那么1千克卵可以在不到一周的时间内转化为 300 千克的蛋白质。

The most costly "vomit" is a substance called ambergris, produced by sperm whales. This substance is used in perfume. It is valued at more than US$ 20 000 per kilogram. This "exquisite" matter is usually discovered by dogs on the beach.

最名贵的"呕吐物"是由抹香鲸形成的一种叫做龙涎香的物质。这种物质被用于制作香水，它的价格高达每千克 2 万余美元。这种宝贝非常容易被海滩边的小狗闻到。

Think About It
想一想

Would you rather take antibiotics, or would you prefer maggots cleaning your wounds?

你是愿意用抗生素，还是愿意让蛆来清理你的伤口？

你认为缓解疼痛的良方是时间，还是止疼药？

Is time a good remedy for pain, or would painkillers be the best option?

Do you think that flies are good friends to human beings?

你认为苍蝇是人类的好朋友吗？

你认为哪个价值更高，是屠宰场的废弃物，还是栖息于废弃物上的蛆的口水？

What has the greatest value, the waste from the abattoir or the spit of the maggots farmed on the waste?

How much protein do you eat? Make a list of all you eat during a normal day. List food in terms of protein, carbohydrates and fats. Now look at what maggots eat, if their main source of nutrition is abattoir waste. Then verify how much of the food you eat created how much waste. And then ask the question: How much waste do the maggots leave behind?

你要食用多少蛋白质呢？把你一天吃过的东西列个清单。将这些食物按照蛋白质、碳水化合物和脂肪来分类。现在，看看蛆都吃什么，它们的主要营养来源都是废弃物。然后，算一下你吃的食物产生了多少废弃物。接下来的问题是，蛆又留下了多少废弃物呢？

TEACHER AND PARENT GUIDE

学科知识
Academic Knowledge

生物学	蛆在治疗过程中所扮演的角色是吞噬细菌；蛆的唾液腺中含有变形杆菌，能产生抑制细菌的物质；苍蝇幼虫能够杀死链球菌，链球菌对于抗生素而言具有很强的抗药性。
化 学	蛆释放减少坏死组织的酶；生物手术的应用随着抗生素的发现而减少，现在细菌变得越来越具有抗药性，苍蝇幼虫治疗法又重新引入；幼虫能够产生碳酸钙来改变细菌繁殖环境；活蛆奶酪是撒丁传统食物，由羊奶制成，里面有活的苍蝇幼虫，能释放强酸来分解奶酪的脂肪，进而做出低脂肪的软奶酪。
物 理	蛆的蠕动刺激抗菌物的释放；基于比重不同来进行物质分离（如在盐水中分离呕吐物，在呕吐物中分离唾液）。
工程学	蛆虫疗法又称生物手术；设计一个口袋让蛆虫施展功效，避免让活的蛆虫直接在伤口上爬动。
经济学	通过比较蛆虫疗法的成本、并不见效的抗生素疗法的成本以及为病人截肢所产生的医疗、社会、经济代价，来估算一下蛆虫疗法的竞争力；饲料成本占畜牧业食品生产成本的60%～70%；替代效应的重要性在于，它不仅能更有效地利用有限的资源，而且还能产生多重效益增加当地经济收入。
伦理学	关于动物权利的争论：是将蛆碾碎来提取活性成分，还是让蛆活着，通过盐水催吐以提取唾液，最后让蛆成为鸡或鹌鹑的食物。
历 史	古埃及人为了食用鹌鹑肉和鹌鹑蛋而饲养鹌鹑；几个世纪以来澳大利亚土著部落一直沿用蛆虫疗法，安第斯山脉的玛雅人也是如此。
地 理	鹌鹑从非洲南部到欧洲的迁徙路径。
数 学	在将产业集群的利润率与核心产业对比时电子数据表格的局限性；多重效益的产生为当地经济催生了更多重的效益。
生活方式	可以通过改善生活方式（如戒烟）和改变饮食习惯（如选择无糖食物）来使伤口更好地愈合；鹌鹑蛋要比鸡蛋营养丰富得多。
社会学	社群中的成员有可能会几乎同时呕吐；秘鲁亚马逊流域的死藤水可导致剧烈呕吐，用于清除胃肠中的寄生虫。
心理学	时间不能治愈所有伤口；我们为什么不喜欢我们不了解的事物；蛆虫治疗法最主要的障碍是心理因素，这种障碍同时存在于患者和医务人员中。
系统论	多元价值的创造并非都能被量化，如卫生条件的改善和减少毒素的摄入。

教师与家长指南

情感智慧
Emotional Intelligence

鹌鹑

鹌鹑对自己的食物很满意，不解人类为何不喜欢这种营养丰富的食物。人类因自己不了解蛆的价值而不喜欢蛆，而鹌鹑却并非这样无知，因此，鹌鹑对人类没有太多认同。鹌鹑认为人类有些无知，而且还不愿意更新自己的知识。鹌鹑有时间观念，享用大自然给他们的充足食物。他们表达了对苍蝇幼虫生产能力的钦佩，并称赞他们奉为世界上最棒的。鹌鹑的对话深入到了超越食物本身的层面，比如关于蛆对医疗卫生方面作用的讨论，这些充满探索的交流覆盖了广泛的议题。总之，这些对话展现了他们的处世哲学和反思性思维，坚持超越自我世界，认为有些东西虽然不被认可，但却是很有用的。

艺术
The Arts

图表为我们理解复杂数据提供了简单的方式。自己制作一张图表吧，数据是屠宰场的废物量，蝇卵和蛆的数量，转化为鹌鹑饲料、人类医疗用品、鱼饲料的数量，海洋中被捕获用于人工繁殖的鱼苗的数量。看看你可以使用哪些不同类型的图表来展现以上数据(如饼图、条状图、柱状图)，然后开动脑筋吧！

TEACHER AND PARENT GUIDE

思维拓展
Systems: Making the Connections

蛆是很多鱼类和鸟类的理想食物，人类却不屑一顾，还担心苍蝇在我们的食物中产卵。然而，蛆是生物链中十分重要的环节，蛆有能力将腐败物的废渣转化成50%~60%的蛋白质和25%的脂肪，这是多么理想的饲料。现在的许多动物饲料都是由大豆和鱼粉组成，可是，大豆仅含有35%的蛋白质，鱼粉的蛋白质含量相对多一些。之所以流失了这么多蛋白质，意谓着蛋白质被用于生产更多的鱼（或鸡）。我们是在将一种食物转化成另一种质量更低的食物，其营养含量远不如最原始的饲料，这也解释了人们为什么要忍受饥饿。现在，我们不仅要关心食物链中损失的营养，还要关心我们可持续发展实践中损失的更好机遇，比如养殖蛆虫。昆虫饲养行业的兴起是一大进步，但我们仍然过于聚焦于单一的行业。那些将蛆虫制作成食物的企业没有看到蛆虫唾液的价值，医疗卫生行业也没有看到提供更健康饮食的机遇。创造成功的关键是实现多赢。创造经济、社会、环境共赢的能力将最终实现可持续的健康发展和食品供应，甚至能为增强土壤肥力提供坚实保障。

动手能力
Capacity to Implement

做一次蛆虫营销大师吧！你知道患者和医护人员都不喜欢蛆，人们也不喜欢接近苍蝇，看到蛆在腐烂的尸体上爬来爬去也会感到恶心。那么，现在你就有责任为蛆虫开展营销工作啦！你要将蛆虫以人类朋友的形象展示出来，告诉人们蛆虫可以为提升我们的生命质量做出很大的贡献。因此，你要想出能够帮助人们克服一知半解和厌恶心理的方法，找到兴趣点，甚至可以让人们变得喜欢苍蝇、蝇卵和蛆虫。不要太严肃，记得要幽默些哦！

教师与家长指南

故事灵感来自

戈弗雷·扎木略
Godfrey Nzamujo

戈弗雷·扎木略出生在尼日利亚北部城市卡诺，后被送到美国加利福尼亚学习。他是一位成功的学者，拥有农学、经济学和信息技术学位。然而，他的理想是改善非洲人民的生活。他放弃了美国的事业返回非洲，在贝宁建立了宋海组织。他在非洲的事业开始于1985年，此前尼日利亚拒绝为他提供可利用的土地。他将自己的全部精力投入到建立一个生产中心，在那里，赤贫及没有受过教育的人都有生产粮食、找到工作的机会。他让一名铁匠制造生产工具，训练农民育种、沤肥。正是在这种充分获取一切可利用之物效益的探索中，扎木略将蛆虫养殖整合为更有价值的产业链。从在波多诺伏的第一个生产中心，扩大到帕罗库、萨瓦罗、肯维基三个地区。宋海组织每年培训300名非洲人，教他们如何生产粮食和获得工作，通过利用现有之物来脱贫致富。

更多资讯

www.slideshare.net/luafiro/terapia-larva-ly-presentacion-de-pacientes

www.lrrd.org/lrrd20/12/anie20205.htm

www.songhai.org/

图书在版编目（CIP）数据

蛆的口水：汉英对照 /（比）鲍利著；田烁译. -- 上海：
学林出版社，2015.6
（冈特生态童书. 第 2 辑）
ISBN 978-7-5486-0866-0

Ⅰ．①蛆… Ⅱ．①鲍… ②田… Ⅲ．①生态环境－环境保护－
儿童读物－汉、英 Ⅳ．① X171.1-49

中国版本图书馆 CIP 数据核字 (2015) 第 086051 号

© 2015 Gunter Pauli
著作权合同登记号 图字 09-2015-446 号

冈特生态童书
蛆的口水

作　　者——	冈特·鲍利	
译　　者——	田　烁	
策　　划——	匡志强	
责任编辑——	李晓梅	
装帧设计——	魏　来	
出　　版——	上海世纪出版股份有限公司 学林出版社	
	地　址：上海钦州南路 81 号　电　话 / 传真：021-64515005	
	网　址：www.xuelinpress.com	
发　　行——	上海世纪出版股份有限公司发行中心	
	（上海福建中路 193 号　网址：www.ewen.co）	
印　　刷——	上海图宇印刷有限公司	
开　　本——	710×1020　1/16	
印　　张——	2	
字　　数——	5 万	
版　　次——	2015 年 6 月第 1 版	
	2015 年 6 月第 1 次印刷	
书　　号——	ISBN 978-7-5486-0866-0/G·315	
定　　价——	10.00 元	

（如发生印刷、装订质量问题，读者可向工厂调换）